Learn 2 Think

200 Challenging Math Problems

Every 1st Grader Should Know

This book belongs to:

_____

Grade:_____

200 Challenging Math Problems

every 1st grader should know

New edition 2017
Copyright Learn 2 Think Pte. Ltd.

Published by:
Learn 2 Think Pte. Ltd.

ISBN: 978-981-07-2762-8

Master Grade 1 Math Problems

Introduction:

Solving math problems is core to understanding math concepts. When Math problems are presented as real-life problems students get a chance to apply their Math knowledge and concepts they have learnt. Word problems progressively develop a student's ability to visualize and logically interpret Mathematical situations.

This book provides numerous opportunities to students to practice their math skills and develop their confidence of being a lifelong problem solver. The multi-step problem solving exercises in the book involve several math concepts. Student will learn more from these problems solving exercises than doing ten worksheets on the same math concepts. The book is divided into 8 chapters. The last chapter of the book explains step wise solutions to all the problems to reinforce learning and better understanding.

How to use the book:

Here is a suggested plan that will help you to crack every problem in this book and outside.

Follow these 4 steps and all the Math problems will be a NO PROBLEM!

Read the problem carefully:

- ✎ What do I need to find out?
- ✎ What math operation is needed to solve the problem? For example addition, subtraction, multiplication, division etc.
- ✎ What clues and information do I have?
- ✎ What are the key words like sum, difference, product, perimeter, area, etc.?
- ✎ Which is the non-essential information?

Decide a plan

- ✎ Develop a plan based on the information that you have to solve the problem. Consider various strategies of problem solving:
- ✎ Drawing a model or picture
- ✎ Making a list
- ✎ Looking for pattern
- ✎ Working backwards
- ✎ Guessing and checking
- ✎ Using logical reasoning

Solve the problem:

Carry out the plan using the Math operation or formula you choose to find the answer.

Check your answer

- ✎ Check if the answer looks reasonable
- ✎ Work the problem again with the answer
- ✎ Remember the units of measure with the answer such as feet, inches, meter etc.

Master Grade 1 Math Problems

Note to the Teachers and Parents:

✎ Help students become great problem solvers by modelling a systematic approach to solve problems. Display the 'Four step plan of problem solving' for students to refer to while working independently or in groups.

✎ Emphasise on some key points:

✎ Enable students to enjoy the process of problem solving rather than being too focused on finding the answers.

✎ Provide opportunities to the students to think; explain and interpret the problem.

✎ Lead the student or the group to come up with the right strategy to solve the problem.

✎ Discuss the importance of showing steps of their work and checking their answers.

✎ Explore more than one possible solution to the problems.

✎ Give a chance to the students to present their work.

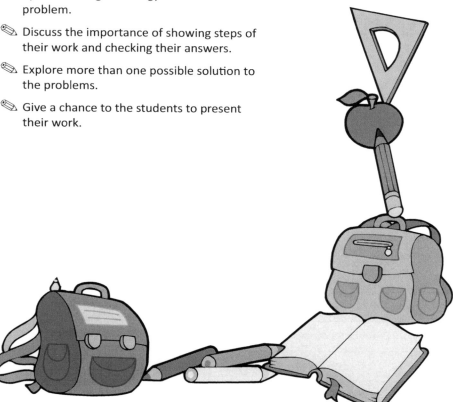

Contents

PROBLEM 1

What is the biggest and the smallest whole number you can make using the digits 7 and 3?

Answer: ....

PROBLEM 2

9 + 2 is less than I am. 8 + 5 is greater than I am. What number am I?

Answer: ....

James is thinking of a 2-digit number. The digit in the tens place is 1 less than 3. The digit in the ones place is 2 more than the digit in the tens place. What number is James thinking of?

Answer: ....

PROBLEM 4

When 14 is added to a number,
the result is 2 more than 32.
What is the number?

Answer: ....

PROBLEM 5

When two numbers are added together, the answer is 30. If one of the numbers is 12, what is the other number?

Answer: ....

PROBLEM 6

I am more than 13 but less than 16. I am not 15. What number am I?

Answer: ....

PROBLEM 7

I am thinking of 2 numbers. If I add them, the answer is 20. If I subtract them, the answer is 4. What are the two numbers?

Answer: ....

PROBLEM 8

Mira has 5 letters. Each letter represents some points.
Mira made a word "HOME".
How many points will she get altogether?

O	M	L	E	H
4	2	5	3	4

Answer: …………………………

PROBLEM 9

Sam used some matchsticks to make two identical squares as shown below. If he wants to add 2 more squares, how many more matchsticks does he need?

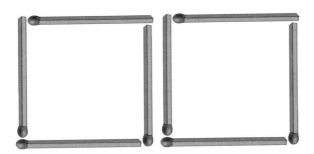

Answer: ....

PROBLEM 10

I am a 2-digit odd number. I am more than 25 but less than 40. The sum of my two-digits number is 8. What number am I?

Answer: ....

PROBLEM 11

Which is the largest two digit number? Also write the smallest two digit number.

Answer: ....

I am a number greater than 9
but less than 19. I am an odd
number. When you count by 5's,
you say my name. What number
am I?

Answer: ....

PROBLEM 13

Write the numbers 0 to 20.
How many 2s did you use?

Answer: ....

The sum of three numbers is 15. The third number is 4 less than the second number. If the second number is 6 what are the first and the third numbers.

Answer: ....

12 more than an unknown number is 42. What is the unknown number?

Answer: ....

PROBLEM 16

The sum of two numbers is 90. The bigger number is 30 more than the smaller number. What are the two numbers?

Answer: ....

What is the largest number you can make only using the digits 2, 8, and 5? You have to use all three of these numbers only once. What is the smallest number you can make using the three digits only once?

Answer: ....

Some children queue up to buy snacks in the school canteen. Anna is first in the queue.

Front | | | | | Back

Anna	Betty	Cathy	Jane	Eve	Lynn
1st					

a) Who is the child just after the 2nd child in the queue?

b) Who is 3rd in the queue after Anna and Betty leave the queue?

Answer: ....

During a race, right before the finishing line, I passed the runner who won the fifth place. What place did I win?

Answer: ....

PROBLEM 20

Read the clues given below and name the trees in correct order as shown in the picture.

Tree A is shorter than Tree B.

Tree C is the tallest.

Label the boxes correctly.

| Tree | Tree | Tree |

Answer: …………………………

PROBLEM 21

Some students line up to board the school bus. John is 2nd in the queue, Maggie is in 4th place. Henry is standing between John and Maggie ? What is Henry's position in the queue?

Answer: ....

PROBLEM 22

Linda is standing in a queue. There are 5 people standing in front of her. What is Linda's place in the queue after two people in front of her leave?

Answer: ....

PROBLEM 23

10 fruits are placed in a row. An orange is the 3rd fruit from the right. At what place is the orange from the left?

Answer: …………………………

PROBLEM 24

Peter sits in the middle of a row. There are 6 boys to his left and 6 boys to his right. How many boys are there in the row?

Answer: ....

Rosy is standing in a queue to buy movie tickets. She is 5th from the front and there are 6 people behind her. How many people are standing in the queue?

Answer: ....

PROBLEM 26

Eva has 4 apples. Jonathan has 5 more apples than Eva. How many apples do they have altogether?

Answer: ....

PROBLEM 27

Gary has 12 apples. Jane has 10 apples. How many apples do they have altogether?

Answer: ………………………….

Lawrence has 20 stamps. Kevin has 12 more stamps than him. How many stamps do they have altogether?

Answer: ....

PROBLEM 29

James had some oranges. He sold 22 of them in the morning and 12 of them in the afternoon. He had 10 oranges left. How many oranges did he have at first?

Answer: ………………………

There were 7 boys and 10 girls in a soccer club, and 16 students in the table tennis club. How many students were there altogether?

Answer: ....

Henry had some mangoes. He sold 6 of them and gave 12 mangoes to his brother. He then had 19 mangoes left. How many mangoes did he have at first?

Answer: ……………………………

Kelvin has 30 stamps. James has 14 more stamps than him. How many stamps do they have altogether?

Answer: ....

Jonathan and Jerry have a total of 40 marbles. Jerry gave 5 marbles to Jonathan and now both of them have the same number of marbles. How many marbles did Jerry have in the beginning?

Answer: ....

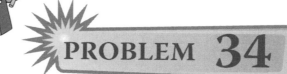

PROBLEM 34

Jennifer had some fish. She sold 25 fish and had 12 left. How many fish did she have at first?

Answer: ....

Jenny bought 22 yellow ribbons and 10 more green ribbons than yellow ribbons. Sharon bought 15 ribbons more than Jenny. How many ribbons did Sharon buy?

Answer: ....

There are 12 green balls in a basket. Henry puts 14 red balls and 22 purple balls more in the basket. How many balls are there in the basket altogether?

Answer: ....

A farmer sold 14 apples, 30 mangoes, and 18 bananas. How many fruits did he sell altogether?

Answer: ....

William baked 16 cakes. He puts them equally into 4 boxes. How many cakes are there in each box?

Answer: ……………………………

There are 7 boys in the table -tennis club. Each boy has 4 table-tennis rackets. How many table tennis rackets do they have altogether?

Answer: ....

PROBLEM 40

Pauline bought 5 bags of papayas. Each bag has 3 papayas. How many papayas did Pauline buy in all?

Answer: ....

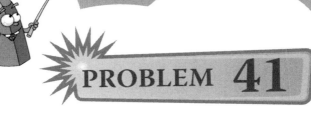

PROBLEM 41

There are 3 red tulips and 4 blue tulips in a vase. Each tulip has 8 petals. How many petals are there altogether?

Answer: ....

PROBLEM 42

On Sunday, there were 12 apricots in a basket. 14 more apricots were put into the basket on Monday. 24 more apricots were put into the basket on Tuesday. How many apricots were there by the end of three days?

Answer: ....

Joe reads 4 pages of a book each day. How many days will he take to finish reading a 12 page book?

Answer: …………………………

PROBLEM 44

There are 3 red pens and 2 blue pens in a box. How many pens are there in 5 such boxes?

Answer: ....

Peter drinks 2 glasses of milk each day. How many glasses of milk does he drink in a week?

Answer: ....

PROBLEM 46

Kevin bought a dozen eggs. He puts them equally into 4 bags. How many eggs were there in each bag?

Answer: ....

There were 10 blue crayons and 6 purple crayons in a box. The crayons were shared equally among 2 children. How many crayons did each child get?

Answer: ....

PROBLEM 48

Bryan had 16 apples. He shared the apples equally with his 2 friends. How many apples did each of them get?

Answer: ....

54

PROBLEM 49

Ben had 8 coins. Lucy had 4 coins. They divided their coins equally among themselves. How many coins did each of them get?

Answer: ……………………

PROBLEM **50**

Jack sold 34 charity tickets. He
sold 12 charity tickets fewer than
Mary. How many charity tickets
did they sell altogether?

Answer: …………………………

PROBLEM 51

Brandon bought some eggs. He ate 12 of them, threw away 6 rotten ones and packed the rest into 3 bags. Each bag had 5 eggs. How many eggs did Brandon have at first?

Answer: …………………………

Gary bought 5 meters of cloth. Jason bought 6 meters more of cloth than Gary. How much cloth did both of them buy altogether?

Answer: ………………………….

PROBLEM 53

Anna baked 20 cookies. Mia baked double the number of cookies than Anna did. How many cookies did they bake altogether?

Answer: ....

PROBLEM 54

Jennifer and Peggy have a total of 6 apples. Jennifer has twice as many apples as Peggy. How many apples does Jennifer have?

Answer: ....

PROBLEM 55

After giving away 2 apples, John has 6 apples left. How many apples did he have in the beginning?

Answer: …………………………

David has 4 pens. He has 7 more
rulers than pens. How many
rulers does David have?

Answer: …………………………

PROBLEM 57

Jake and Peter went fishing.
Jake caught 5 fish. Peter caught
3 more fish than Jake. How
many fish did Peter catch?

Answer: ....

PROBLEM 58

Jason bought 32 tomatoes from the supermarket. Kelvin bought 10 tomatoes less than Jason. How many tomatoes did Kelvin buy?

Answer: …………………………

Paul bought 26 chocolates.
He gave 15 chocolates to his
daughter. How many
chocolates did he have left?

Answer: ....

PROBLEM 60

Kelly picked 7 pears. Nina picked 5 pears. How many more pears did Kelly pick than Nina?

Answer: …………………………

Tim bought 16 tickets for a movie. He gave 4 tickets to his friends and used the rest of the tickets to take his family to the movie. How many people are there in Tim's family?

Answer: ....

PROBLEM 62

Alison had some marbles. After her mother gave her 18 marbles, she had 30 marbles. How many marbles did she have at first?

Answer: ....

PROBLEM 63

There were 72 people at a party. 20 of them were children and the rest were adults. How many more adults than children were there at the party?

Answer: ....

Jason has 28 bananas. Jessica has 52 bananas. How many more bananas should Jason buy so that he has the same number of bananas as Jessica?

Answer: …………………………

When a number is subtracted from another, the answer is 13. If the bigger number is 18, what is the smaller number?

Answer: ....

PROBLEM 66

There are 9 lamps in my house. If 2 lamps are turned off how many lamps are still lit?

Answer: ....

PROBLEM 67

There were 38 soccer players practicing at a stadium. 15 of them wore green jerseys and the rest wore red. How many more players wore red jerseys than green jerseys?

Answer: ....

PROBLEM 68

Colin has 34 pencils. He has 18 more pencils than erasers. How many erasers does Colin have?

Answer: …………………………

PROBLEM 69

Nancy has 10 apples. Grace has 6 apples. How many apples should Nancy give to Grace so that they both have an equal number of apples?

Answer: ....

PROBLEM 70

Carol's house has 9 bulbs but three of the bulbs do not work. How many bulbs are working in Carol's house?

Answer: ....

PROBLEM 71

Mary bought 12 chicken pies. Ali bought 8 chicken pies. How many more chicken pies did Mary buy than Ali?

Answer: ....

A book has 40 pages. Melanie has read 15 pages of the book. How many more pages does she need to read to finish the book?

Answer: …………………………

Kelvin and Jonathan have a total of 9 coins. Jonathan has 3 coins more than Kelvin. How many coins does Kelvin have?

Answer: ………………………………

PROBLEM 74

Maria needs 19 pineapples for making pineapple juice. She already has 17 pineapples. How many more pineapples does she need to buy to make the juice?

Answer: ....

Alvin has 20 balloons. He has 9 balloons more than Ross. How many balloons does Ross have?

Answer: ....

PROBLEM 76

Karen planted 13 carrot seeds. Only 5 of them sprouted. How many carrot seeds did not sprout?

Answer: ....

Fifteen children are attending a birthday party. Nine of them are girls. How many boys are attending the party?

Answer: ....

PROBLEM **78**

Farmer Jim had a total of 18 chicken and ducks. If he had 10 chicken, how many ducks did he have?

Answer: ....

PROBLEM 79

Kim's school is 13 kilometers away from her home. She has already walked 4 kilometers out of that. How far does she still have to go to reach the school?

Answer: ....

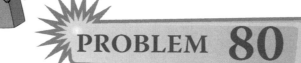

PROBLEM 80

Mary has 45 meters of ribbon. She gave 10 meters from it to her daughter and 10 meters to her friend Grace. What is the length of the remaining piece of ribbon?

Answer: ………………………….

There are 30 roses in a vase. 7 of them are red, 4 of them are yellow and the rest are pink. How many pink roses are there in the vase?

Answer: ....

Lucy baked a total of 35 cupcakes. She baked 9 coffee cupcakes, some strawberry cupcakes and 18 banana cupcakes. How many strawberry cupcakes did Lucy bake?

Answer: ....

Vanessa baked 19 lemon cakes and 26 chocolate cakes on Monday. On Tuesday, she baked 15 chocolate cakes. How many more chocolate cakes than lemon cakes did Vanessa bake?

Answer: ....

PROBLEM 84

There are 21 boys and 12 girls in the table-tennis club. There are 16 boys and 22 girls in the chess club. How many fewer girls than boys are there in both the clubs?

Answer: ……………………………

There are 35 green candies and 12 red candies in a jar. After Thomas ate some of the candies, there were 16 candies left. How many candies did Thomas eat?

Answer: ....

Jason has a collection of some stamps. He received 16 stamps from his father and 12 stamps from his brother. He had 50 stamps in the end. How many stamps did he have in the beginning?

Answer: ....

PROBLEM 87

There were 35 adults and 28 children at a party. After 4 men and 18 girls left, how many people were there at the party ?

Answer: ....

PROBLEM 88

Mrs Watson had 32 eggs .She threw away 10 rotten eggs, sold away 6 and kept the rest with her. How many eggs did Mrs Watson keep?

Answer: ....

PROBLEM 89

There are 52 pages in a magazine. Janice read 12 pages on Tuesday and 6 pages on Wednesday. How many pages has she not read yet?

Answer: ....

PROBLEM 90

There are 16 boys and 24 girls at the school playground. Their P.E teacher wants to give two balls to each of them but she has only 40 balls with her right now. How many more balls does she need for everyone to have 2 balls?

Answer: ....

PROBLEM 91

8 students went to the library to borrow books. 2 of them did not not borrow any books. The rest of the students borrowed 3 books each. How many books did all of them borrow altogether?

Answer: ………………………

PROBLEM 92

James scored 40 points in a game. Jason scored 8 points lesser than him. How many points did they score altogether?

Answer: ....

There were 28 paper clips in a box. Kelvin had 2 such boxes. He used 8 paper clips and gave away 3 to his friend. How many paper clips were left in the box?

Answer: ....

Jessica bought 5 packets of balloons that had 10 balloons in each packet. She used 6 balloons out of them. How many balloons did Jessica still have left?

Answer: ....

Jason, Anna and Tom had 80 marbles in all. Jason had 25 marbles. Anna had 12 marbles more than him. How many marbles did Tom have?

Answer: ....

There are 38 roses in a vase. 12 of them are red, 10 are pink and the rest are white. How many white roses are there in the vase?

Answer: ....

PROBLEM 97

Mr. Smith wants to buy 60 fruits. He bought 18 apples and 19 bananas from the supermarket. How many more fruit does Mr. Smith need to buy?

Answer: ....

PROBLEM 98

Box A contains 52 cards. Box B contains twice as many cards as Box A. How many cards are there in the two boxes?

Answer: …………………………

There are 26 boys and 14 girls in a class. 17 children wear spectacles. How many children do not wear spectacles?

Answer: ....

PROBLEM 100

Christina had 80 cookies. She gave 7 cookies each to her 3 friends. How many cookies did she have left?

Answer: ....

Jane had 35 stickers. She gave 12 stickers to Samuel. William gave her 3 more stickers to Jane. How many stickers did Jane have in the end?

Answer: ....

PROBLEM 102

There are 45 books on a book shelf. 12 of them are English books, 7 are science books and the rest are math books. How many math books are there on the shelf?

Answer: ……………………………

PROBLEM 103

There were 54 apples in a basket.
There were 12 fewer oranges
than apples in the same basket.
How many fruits were there
altogether in the basket?

Answer: ....

There were 40 fishes in a fish tank. 16 of them were orange, 11 were red and the rest were white. How many white fish were there in the tank?

Answer: ....

PROBLEM 105

A zoo-keeper had to feed 2 bananas to every monkey. There were 12 monkeys and he had only 20 bananas in his basket. How many more bananas would he need to feed all the monkeys?

Answer: …………………………

PROBLEM 106

There are some children sitting on a row of chairs. No chair is empty. The 1st chair starts from the right. Sam sits on the 5th chair. Ricky sits on the 3rd chair from the left. There are 4 children sitting between Ricky and Sam. How many children are there altogether?

Answer: ....

PROBLEM 107

Joel has a toy car collection. His mother gave him 18 toy cars more and his father gave him another 8 toy cars. He now has a total of 40 toy cars in his collection. How many toy cars did Joel have at first?

Answer: ....

Thomas had 28 toy soldiers. 12 of them were red, 5 were green and the rest were blue. How many more blue toy soldiers than green toy soldiers did he have?

Answer: ....

80 students participated in a chess competition. 15 of them were in Grade 2, 13 were in Grade 3 and the rest were in Grade 4. What was the total number of Grade 3 and Grade 4 students who participated in the competition?

Answer: ....

PROBLEM 110

There are 46 tulips in a vase. There are 28 more tulips than roses in the vase. What is the total number of flowers in the vase?

Answer: ....

A photo album contained 25 photographs. Jason took out 6 photographs and put in 12 new ones. How many photographs are there in the photo album now?

Answer: ....

PROBLEM 112

Susan baked 7 strawberry cupcakes. Claire baked 3 fewer cupcakes than Susan. How many strawberry cupcakes did they bake altogether?

Answer: ………………………

PROBLEM 113

Jason and Kevin have 25 books. Jason has 5 more books than Kevin. How many books does Jason have?

Answer: ………………………

PROBLEM 114

Larry had 16 oranges. He gave 7
of them to his friend and gave 2
to his mother. How many
oranges was Larry left with?

Answer: …………………………

PROBLEM 115

Samuel and Jason have a total of 40 fish. Samuel has 10 fish more than Jason. How many fish does Samuel have?

Answer: ....

PROBLEM 116

Liz and Kevin have a total of 50 eggs. Kevin has 10 eggs more than Liz. How many eggs should Kevin give to Liz so that both of them have the same number of eggs?

Answer: …………………………

PROBLEM 117

Tom has 10 marbles. James has 7 marbles fewer than Tom. How many marbles do they have altogether?

Answer: ....

PROBLEM 118

Jake has 14 marbles. He lost 4 marbles while playing. He bought 6 new marbles. How many marbles does Jake have now?

Answer: …………………………

PROBLEM 119

Alex has 200 balloons. 3 of them are red, 45 of them are blue and the rest are yellow. How may yellow balloons does Alex have?

Answer: ………………………

PROBLEM 120

There are 100 students in a school. There are 30 boys and 25 girls and the rest were teachers. How many more teachers than girls were there?

Answer: ....

Maya and her family went on a vacation for 14 days. They spent 5 days in Germany and 4 days in Italy. They spent rest of the time in France. How many days did they spend in France?

Answer: ....

PROBLEM 122

There are 30 books on a bookshelf. 10 of them are English books, 15 are Science books and the rest are Math books. How many Math books are there?

Answer: ....

PROBLEM 123

Lucy baked 30 cupcakes. She baked 10 coffee cupcakes, some strawberry cupcakes and 10 banana cupcakes. How many strawberry cupcakes did she bake?

Answer: ....

Mum wants to give 20 oranges to Amie, Nathan, Mike and Daniel. How many oranges does each child receive?

Answer: ....

PROBLEM 125

On a farm, there are 5 sheep, 5 ducks, 2 cows, 2 cats and the farmer. How many legs are there altogether?

Answer: …………………………

PROBLEM 126

Alice put 8 stamps on every page of her stamps album. There were 5 pages in her album. How many stamps did she have altogether?

Answer: …………………………

PROBLEM 127

There were 20 marbles in a bag. Sam divided them equally into 4 groups. How many marbles were there in 2 such groups?

Answer: ....

PROBLEM 128

There were 12 chicken eggs and 6 duck eggs in a basket. All the eggs were packed equally into 3 boxes. How many eggs would there be in each box?

Answer: ....

PROBLEM 129

Bryan had 4 bags. There were 9 oranges in each bag. He gave 12 oranges to his friend. How many oranges did he have left?

Answer: …………………………

PROBLEM 130

Richard had 3 boxes of cookies. There were 5 cookies in each box. His friend gave him another 12 cookies. How many cookies did Richard have altogether?

Answer: …………………………

PROBLEM 131

Karen packed 12 buns into 3 packets. In each packet, there was 1 raisin bun and some butter buns. How many butter buns were there in each packet?

Answer: …………………………

PROBLEM 132

There were 6 flowers in a bouquet. Dorothy bought 3 such bouquets. How many flowers did she buy altogether?

Answer: ....

There are 4 bags of oranges. Each bag has 9 oranges. There are 3 bags of mangoes. Each bag has 6 mangoes. How many fruits are there altogether?

Answer: ………………………

PROBLEM 134

16 buns are shared equally among 4 children. Each child eats 2 buns. How many buns are left in the end?

Answer: ………………………….

PROBLEM 135

2 books cost $10 and 4 pens cost $24. How much does Henry pay if he buys one book and one pen?

Answer: …………………………

A fruit seller has 8 bunches of bananas. Each bunch has 4 bananas. 23 of the bananas are rotten, and the rest of the bananas get sold. How many bananas did the fruit seller sell?

Answer: …………………………

Kelvin has a total of 16 potatoes. He wants to put 4 potatoes in one bag. How many bags does he need to pack all the potatoes?

Answer: …………………………

A teacher asked her students to stand in 3 rows that had 7 students each. After some time 2 new students were added to each row. How many students were there now altogether.

Answer: …………………………

Thomas had 10 boxes of egg tarts. There were 4 egg tarts in each box. He gave 6 boxes of egg tarts to his friends. How many egg tarts was Thomas left with?

Answer: ....

Larry bought 30 lemons. He wants to pack all the lemons into packets of 5. How many packets would he need to pack all the lemons?

Answer: ....

Richard has 35 bananas. He wants to pack 5 bananas into one packet and sell each at $2.

a) How many packets did Richard pack?

b) How much money did he earn after selling all the bananas?

Answer: ……………………………

PROBLEM 142

Jeffrey has 2 mangoes. Harry has twice the amount of mangoes than Jeffrey. Victor has thrice the amount of mangoes than Jeffrey. How many mangoes do the 3 children have altogether?

Answer: ………………………

PROBLEM 143

A chicken has 2 legs, and a dog has 4 legs. There were 4 chickens and 2 dogs on the farm. How many legs were there in all ?

Answer: ....

PROBLEM **144**

There were 4 pigs and some birds on the farm. There were a total of 20 legs. How many birds were there on the farm?

Answer: …………………………

PROBLEM 145

Ella has 15 stickers. She gives
3 stickers to each of her 3 sisters.
How many stickers does she have
left?

Answer: ………………………….

PROBLEM 146

Lena has 15 cookies. She packs 5 cookies in each box. How many boxes does she need to pack 15 cookies?

Answer: ....

PROBLEM **147**

Tom has 3 cards now. He buys 3 new cards every day. How many days would it take Tom to collect 18 cards altogether?

Answer: …………………………

There are 4 sparrows sitting on the branch of a tree. How many feet are there on the branch?

Answer: ....

Mum baked some cupcakes for Tim, Liz, and Jill. Each of them ate 2 cupcakes and none were left. How many cupcakes did Mum bake?

Answer: ……………………………

Alice drew 2 flowers on every page of her drawing book. There were 10 pages in her drawing book. How many flowers did she draw altogether?

Answer: …………………………

PROBLEM 151

Tina's birthday is on November 21st. Lucy's birthday is a week before that. When is Lucy's birthday?

Answer: …………………………

PROBLEM 152

Today is May 4th and your birthday is on May 21st. How many more days would you have to wait for your birthday to come?

Answer: ....

PROBLEM 153

Use Ben's schedule to draw hands on the clock.

Ben's Afternoon Schedule

Time	Activity
2:00	Leave School
2:30	Eat Snack
3:00	Play outside
4:00	Do homework

1. Eat snack

2. Do homework

3. What does Ben do at 3:00?

Answer: ……………………………

PROBLEM 154

Eva is 4 years old and Jane is 8. How old will Eva be when Jane is 11years old?

Answer: ....

PROBLEM 155

The sum of Kelvin and his father's age is 50 years. His father is 40 years older than him. Find the age of Kelvin and his father.

Answer: ....

PROBLEM 156

Ann is twice as old as Bill. Bill is three times as old as Carol. If Bill is 3 years old, how old are the 3 children?

Answer: ....

PROBLEM 157

Andy reads two pages every five minutes. How many pages would Andy have read after twenty minutes?

Answer: …………………………

PROBLEM 158

Kevin wakes up at 6 o'clock every day. If he slept at 10 P.M. in the night, how many hours did he sleep?

Answer: ………………………

164

PROBLEM 159

A movie started at 12P.M. and ended after 3 hours. What time did the movie end?

Answer: ....

Carl attended a birthday party.
The party started at 2.00 P.M.
and ended at 4.00 P.M. How
long did the birthday party last?

Answer: …………………………

PROBLEM 161

Tom left for work at 8:00 A.M. He got stuck in the traffic for 2 hours. What time did Tom reach at work?

Answer: ....

PROBLEM 162

Mike does his homework one hour after his dinner. If he eats his dinner at 6:30 P.M., at what time does he start doing his homework?

Answer: …………………………

PROBLEM 163

Lucy started baking a cake at 1:00 P.M. and it should bake for 3 hours. At what time will the cake be ready?

Answer: ....

PROBLEM 164

I went cycling to the park yesterday. It usually takes me 3 hours to reach there. Yesterday I left from home at 9:00 A.M. What time did I reach the park?

Answer: ....

PROBLEM **165**

Danny had a total of 6 apples and oranges. There were 2 more oranges than apples. He sold oranges at a price of $1.

a) How many oranges did he have?

b) How much money did he earn by selling the oranges?

Answer: ....

PROBLEM 166

Amelia and Raymond together have $10. Raymond gave $2 to Amelia and they now have the same amount of money. How much money did Raymond have in the beginning?

Answer: ....

Peter had $20. He spent $4 on Monday and $8 on Tuesday. How much money did he have left?

Answer: ....

PROBLEM 168

Jake saved his money for a whole month. In week 1 he saved $10. In week 2 he saved $15. In week 3 he saved $8. And in week 4 he saved $12. How much money did he save that month?

Answer: …………………………

A watch and a clock cost $59 in a sale. The watch cost $39. How much more did the watch cost than the clock?

Answer: ……………………………

PROBLEM 170

Peter had 90 cents in his wallet. He had 5 ten-cent coins and some twenty-cent coins. How many twenty-cent coins did he have?

Answer: ………………………

A shirt cost $41. A pair of pants cost $12 more than the shirt. How much would the shirt and the pair of pants cost together?

Answer: ………………………

Andy spent $72 on a belt and saved the rest of his money. If he saved $12,

a) How much money did he have at first?

b) How much more did he spend than save?

Answer: ....

Maggie had $15. She bought a chocolate cake and had $7 left. How much would 2 such chocolate cakes cost?

Answer: ………………………

PROBLEM 174

Jason had $80. Linda had $36.
How much more money did
Jason have than Linda?

Answer: ………………………

A book costs $5 more than a magazine. The magazine costs $9. What is the cost of 2 such books?

Answer: ………………………….

PROBLEM 176

Lawrence had $20. He bought a vase for $12. The cashier gave him a change of 1 five dollar note, some two dollar notes and a one dollar note. How many two dollar notes did the cashier give Lawrence?

Answer: …………………………

182

PROBLEM 177

A stalk of rose cost $5. A florist sold 3 stalks of roses and some stalks of orchids. She earned $27. How much money did she earn from selling the orchids?

Answer: ....

Three friends went to have dinner together that cost $100. Ethan paid $35, Marcus paid $40 and Leonard paid the rest. How much did Leonard pay?

Answer: ………………………….

PROBLEM 179

A plate cost $8 more than a bowl. If the plate cost $23, how much did the bowl cost? What was the total cost of the plate and the bowl?

Answer: ....

PROBLEM 180

2 chairs and a table cost $50. A chair costs $8. What is the cost of the table?

Answer: ………………………….

PROBLEM 181

Maggie bought an electric oven that cost $59. She gave $100 to the cashier . How much change did she get back?

Answer: ………………………

PROBLEM 182

Jason had $40. He spent $6 on food, $9 on books and some on stationery and had $8 left. How much did he spend on stationery?

Answer: …………………………

PROBLEM 183

John had $8. Doris had $3 more than John and Paul had $5 less than Doris. How much money did they have altogether?

Answer: …………………………

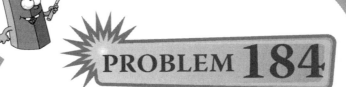

PROBLEM 184

An orange costs 50 cents. A pear costs 20 cents less than an orange. A mango costs 60 cents more than the pear. How much does the pear cost?

Answer: ....

A dress cost $28. It costs $18 more than a blouse. What is the total cost of the dress and blouse?

Answer: …………………………

PROBLEM 186

Richard had $100. He spent $26 and saved the rest of his money. How much money did he save?

Answer: …………………………

PROBLEM 187

A bunch of roses costs $4. There are 3 roses in a bunch. If Caroline spends $20 on buying roses, how many roses does she buy?

Answer: …………………………

PROBLEM 188

Henry bought some biscuits that cost $6. He bought a drink at $8. How much money did he spend altogether?

Answer: ....

PROBLEM 189

Larry had 3 ten-dollar notes that he spent to buy 4 cupcakes at a coffee shop at $5 each. How much change did he get from the cashier?

Answer: ....

PROBLEM 190

John saves $2 every day. Kelvin saves $1 less than John every day.

a) For how many days will John have to save if he wants to save a total of $14?

b) How much will Kelvin save by the time John saves $16?

Answer: ....

Gavin went to a shop with a $10 note. He bought 2 packets of potato chips at $2 each and 2 cans of Coca-Cola at $1 each.

a) How much did Gavin spend on the things he bought?

b) How much change did he receive from the cashier?

Answer: ....

PROBLEM 192

Jenny went to the market with 1 fifty dollar note. She bought 2 fish at $5 each. She bought 3 watermelons at $10 each. How much change did she receive from the cashier?

Answer: …………………………

PROBLEM 193

Kevin bought 6 pairs of socks at $2 each. He gave $20 to the cashier.

a) How much did he spend in all to buy the socks?

b) How much change did he get from the cashier?

Answer: ....

PROBLEM 194

Ben had three $10 notes and two $5 notes with him. His father gave him one $2 note. Using all his money he bought a game for $40. How much money did he have left with him?

Answer: ………………………

PROBLEM 195

Tim paid $5 for a chocolate. He received two 50 cent coins in change. How much did the chocolate cost?

Answer: ....

PROBLEM 196

I want to buy a pencil that costs 50 cents. I pay a 2 dollar note to the shopkeeper. How much change will I get back?

Answer: ....

On the scales shown in the picture, what is the weight of the fruits in the basket?

Answer: ....

PROBLEM 198

The temperature at noon today was 40 degree, but it dropped by 10 degrees in the evening. What was the temperature in the evening?

Answer: ....

I measured my pencil at 9 inches.
My friend's pencil is 16 inches.
How much longer is my friend's
pencil than mine?

Answer: ....

PROBLEM 200

My study table is 5 feet wide. The room's door is 3 feet wide. How much wider does my room's door need to be to fit my desk in?

Answer: ....

206

SOLUTIONS

 ## Solution to Question 1

The biggest and the smallest whole number you can make using the digits 7 and 3 are 73 and 37.

 ## Solution to Question 2

9 + 2 is less than I am.
This means 11 is less than the mystery number.
8 + 5 is greater than I am.
This means 13 is greater than the mystery number.
The number is 12.

 ## Solution to Question 3

James is thinking of a 2-digit number.
The digit in the tens place is 1 less than 3 = 3 − 1 = 2
The digit in the ones place is 2 more than the digit in the tens place = 2 + 2 = 4
The number James is thinking of is 24.

 ## Solution to Question 4

When 14 is added to a number, the result is 2 more than 32.
Result = 32 + 2 = 34
The number is 34 − 14 = 20

 ## Solution to Question 5

When two numbers are added together, the answer is 30.
One of the numbers is 12
The other number = 30 − 12 = 18

208

Solution to Question 6

I am more than 13 but less than 16. I am not 15.
I am 14.

Solution to Question 7

I think of 2 numbers
If I add them the answer is 20. (12 + 8 = 20)
If I subtract them the answer is 4. (12 − 8 = 4)
The two numbers are 12 and 8.

Solution to Question 8

HOME = 4 + 4 + 2 + 3 = 13

Solution to Question 9

Total matchsticks used for 2 identical squares = 8
For 2 more squares, Sam needs 8 more matchsticks.

Solution to Question 10

I am a 2-digit odd number.
I am more than 25 but less than 40. Odd numbers
between 25 and 40 are
27, 29, 31, 33, 35, 37, 39
The sum of my 2-digit number is 8. (3 + 5 = 8)
The number is 35.

Solution to Question 11

The largest two digit number = 99
The smallest two digit number = 10

Solution to Question 12

I am a number greater than 9 and less than 19.
I am an odd number.
Odd numbers between 9 and 19 are 11, 13, 15, 17
When you count by 5's, you say my name. 15 is the only number you say when you count by 5's.
The number is 15.

Solution to Question 13

Total number of 2s used = 3

Solution to Question 14

The sum of three numbers is 15.
The second number is 6.
The third number is 4 less than the second number = 6 − 4 = 2

Sum of second and third number = 6 + 2 = 8
First number = 15 − 8 = 7
First and the third number are 7 and 2.

Solution to Question 15

12 more than an unknown number is 42.
The number is 42 − 12 = 30

Solution to Question 16

The sum of two numbers is 90.
The bigger number is 30 more than the smaller. The bigger number = 30 + 30 = 60
The two numbers are 30 and 60.

Solution to Question 17

Largest number = 852
Smallest number = 258

Solution to Question 18

Child just after the 2nd child is Cathy.
After Anna and Betty leave the queue, the 3rd child is Eve.

Solution to Question 19

During a race, right before the finish line, I passed the runner who won the fifth place. Therefore I won the 4th place.

Solution to Question 20

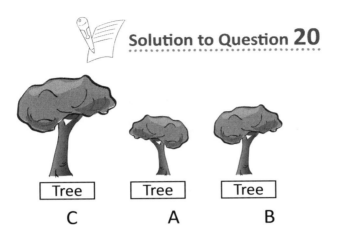

Tree C   Tree A   Tree B

Solution to Question 21

John is 2nd in the queue.
Maggie is in 4th place.
Henry is standing between John and
Maggie. Henry's position in the queue is 3rd.

Solution to Question 22

Linda is standing in a queue.
There are 5 people standing in front of her.
Linda's place = 6th
Linda's place in the line after two people in front of her have left
= 6 − 2 = 4th

 **Solution to Question 23**

Orange is the 8th fruit from left.

 **Solution to Question 24**

Peter sits in the middle of a row.
There are 6 boys to his left and 6 boys to his right.
Number of boys in the row = 6 + 6 + 1 = 13

 **Solution to Question 25**

Rosy is standing in a queue to buy movie tickets.
She is 5th from the front and there are 6 people behind her.
Number of people standing in the queue = 5 + 6 = 11

 **Solution to Question 26**

Eva has 4 apples.
Jonathan has 5 more apples than Eva = 4 + 5 = 9
No. of apples they have altogether = 9 + 4 = 13

 **Solution to Question 27**

Gary has 12 apples.
Jane has 10 apples.
No. of apples they have altogether
= 12 + 10 = 22

Solution to Question 28

Lawrence has 20 stamps.
Kevin has 12 more stamps than him
= 20 + 12 = 32
No. of stamps they have altogether
= 20 + 32 = 52

Solution to Question 29

James had some oranges.
He sold 22 of them in the morning and 12 of them in the afternoon.
Total number of oranges sold = 22 + 12 = 34
He had 10 oranges left.
Number of oranges he had at first = 34 + 10 = 44

Solution to Question 30

There were 7 boys and 10 girls in a soccer club and 16 students in the table tennis club.
Number of students altogether = 7 + 10 + 16 = 33

Solution to Question 31

Henry had some mangoes.
He sold 6 of them and gave 12 mangoes to his brother.
Total number of mangoes = 6 + 12 = 18
He then had 19 mangoes left.
No. of mangoes Henry had at first = 18 + 19 = 37

Solution to Question 32

Kelvin has 30 stamps.
James has 14 more stamps than him
= 30 + 14 = 44
No. of stamps they have altogether =
30 + 44 = 74

Solution to Question 33

Jonathan and Jerry have a total of 40 marbles.
If they had equal number of marbles each would have 20 marbles.
But they have the same number of marbles when Jerry gave 5 marbles to Jonathan. This means Jerry had 5 extra marbles in the beginning.
Number of marbles Jerry had in the beginning
= 20 + 5 = 25

Solution to Question 34

Jennifer had some fish.
She sold 25 fish and had 12 left.
Number of fish she had at first = 25 + 12 =
37

Solution to Question 35

Jenny bought 22 yellow ribbons and 10 more green ribbons than yellow ribbons
Total number of ribbons with Jenny = 22 + 32 = 54
Sharon bought 15 ribbons more than Jenny.
Number of ribbons Sharon bought = 54 + 15 = 69

Solution to Question 36

There are 12 green balls in a basket. Henry
puts 14 red balls and 22 purple balls more
in the basket.
Number of balls in the basket altogether
= 12 + 14 + 22 = 48

Solution to Question 37

A farmer sold 14 apples, 30 mangoes, and 18
bananas. Number of fruits he sold altogether
= 14 + 30 + 18 = 62

Solution to Question 38

William baked 16 cakes .
He puts them equally into 4 boxes.
4 + 4 + 4 + 4 = 16
No. of cakes in each box = 4

Solution to Question 39

There are 7 boys in table -tennis club.
Each boy has 4 table -tennis rackets.
Number of table tennis rackets they have
altogether = 4 + 4 + 4 + 4 + 4 + 4 + 4 = 28

Solution to Question 40

Each bag has 3 papayas.
Pauline bought 5 bags of papayas. Number of
papayas Pauline bought in all

= 3 + 3 + 3 + 3 + 3 = 15

Solution to Question 41

There are 3 red tulips and 4 blue tulips in a vase.
Total number of tulips = 3 + 4 = 7
Each tulip has 8 petals
Number of petals altogether
= 8 + 8 + 8 + 8 + 8 + 8 + 8 = 56

Solution to Question 42

On Sunday, there were 12 apricots in a basket.
14 more apricots were put into the basket on Monday = 12 + 14 = 26
24 more apricots were put into the basket on Tuesday= 26 + 24 = 50
Number of apricots by the end of three days
= 50

Solution to Question 43

Joe reads 4 pages of a book each day
= 4 + 4 + 4 = 12
Number of days he will take to finish reading a 12 page book = 3

Solution to Question 44

There are 3 red pens and 2 blue pens in a box. Number of pens = 3 + 2 = 5
Number of pens in 5 such boxes
= 5 + 5 + 5 + 5 + 5 = 25

 **Solution to Question 45**

Peter drinks 2 glasses of milk each day.
1 week = 7 days
Number of glasses of milk he drinks in a
week = 2 + 2 + 2 + 2 + 2 + 2 + 2 = 14

 Solution to Question 46

Kevin boughta dozen eggs.
1 dozen = 12
He puts them equally into 4 bags.
= 3 + 3 + 3 + 3 = 12
Number of eggs in each bag = 3

 Solution to Question 47

There were 10 blue crayons and 6 purple crayons in a box.
Total crayons = 10 + 6 = 16
The crayons were shared equally among 2 children.
= 8 + 8 = 16
Number of crayons each child got = 8

 Solution to Question 48

Bryan had 16 apples.
He shared the apples equally with his 2 friends.
= 8 + 8 = 16
Number of apples each of them got = 8

218

Solution to Question 49

Ben had 8 coins.
Lucy had 4 coins.
Total = 8 + 4 = 12
They divided their coins equally among themselves =
6 + 6 = 12 .
Number of coins each of them got = 6

Solution to Question 50

Jack sold 34 charity tickets.
He sold 12 charity tickets fewer than Mary.
Number of tickets Mary sold = 34 + 12 = 46
Number of charity tickets they sold
altogether = 34 + 46 = 80

Solution to Question 51

Brandon bought some eggs.
He ate 12 of them and threw away 6 rotten ones
Total = 12 + 6 = 18
He then packed the rest into 3 bags.
Each bag had 5 eggs.
Eggs in 3 bags = 5 + 5 + 5 = 15
Number of eggs he had at first = 15 + 18 = 33

Solution to Question 52

Gary bought 5 meters of cloth.
Jason bought 6 meters more cloth than Gary
= 6 + 5 = 11 meters
Total cloth both of them bought altogether = 11 + 5 = 16 meters

Solution to Question 53

Anna baked 20 cookies.
Greg baked double the number of cookies than Anna did = 20 + 20 = 40
Number of cookies they baked altogether
= 40 + 20 = 60

 Solution to Question 54

Jennifer and Peggy have a total of 6 apples
Jennifer has twice as many apples as Peggy = 2 + 4 = 6
Number of apples Jennifer has = 4

 Solution to Question 55

After giving away 2 apples, John has 6 apples left
Number of apples John had in the beginning = 2 + 6 = 8

 Solution to Question 56

David has 4 pens.
He has 7 more rulers than pens
Number of rulers David has = 4 + 7 = 11

 Solution to Question 57

Jake caught 5 fish.
Peter caught 3 more fish than
Jake = 5 + 3 = 8 fish

Solution to Question 58

Jason bought 32 tomatoes from the supermarket.
Kelvin bought 10 tomatoes less than Jason.
Number of tomatoes Kelvin bought
= 32 − 10 = 22

Solution to Question 59

Paul bought 26 chocolates
He gave 15 chocolates to his daughter.
Number of chocolates he had left = 26 − 15 = 11

Solution to Question 60

Kelly picked 7 pears.
Nina picked 5 pears.
Number of pears Kelly picked more than Nina = 7 − 5 = 2

Solution to Question 61

Tim bought 16 tickets for a movie.
He gave 4 tickets to his friends and used the rest of the tickets to take his family to the movie.
Number of people in Tim's family = 16 − 4 = 12

Solution to Question 62

Alison had some marbles.
After her mother gave her 18 marbles, she had 30 marbles in the end.

Number of marbles she had
at first = 30 − 18 = 12

 ## Solution to Question 63

There were 72 people at a party.
20 of them were children and the rest were adults.
Adults = 72 − 20 = 52
Number of adults more than the number of children
= 52 − 20 = 32

 ## Solution to Question 64

Jason has 28 bananas.
Jessica has 52 bananas.
Number of bananas Jason should buy more so that he has the
same number of bananas as Jessica = 52 − 28 = 24

 ## Solution to Question 65

When a number is subtracted from another, the answer is
13. The bigger number is 18.
The smaller number = 18 − 13 = 5

 ## Solution to Question 66

There are 9 lamps in my house.
2 lamps are turned off.
Number of lamps that are still lit = 9 − 2 = 7

 ## Solution to Question **67**

There were 38 soldiers in a field.
15 of them wore green uniforms and the rest wore brown uniforms. Number of soldiers who wore brown uniforms
= 38 − 15 = 23
Number of soldiers who wore brown uniforms more than the green uniforms = 23 − 15 = 8

 ## Solution to Question **68**

Colin has 34 pencils.
He has 18 more pencils than erasers.
Number of erasers that Colin has = 34 − 18 = 16

 ## Solution to Question **69**

Nancy has 10 apples.
Grace has 6 apples.
Difference = 10 − 6 = 4
Number of apples Nancy must give to Grace so that they have an equal number of apples = 2

 ## Solution to Question **70**

Carol's house has 9 bulbs.
Three bulbs do not work.
Number of bulbs working in Carol's
house = 9 − 3 = 6

 **Solution to Question 71**

Mary bought 12 chicken pies.
Ali bought 8 chicken pies..
Number of chicken pies Mary bought more than Ali = 12 − 8 = 4

 **Solution to Question 72**

A book has 40 pages.
Melanie has read 15 pages of the book.
Number of pages she needs to read more to finish the book =
40 − 15 = 25

 **Solution to Question 73**

Kelvin and Jonathan have a total of 9 coins.
Jonathan has 3 coins more than Kelvin
= 9 − 3 = 6
Number of coins Kelvin has = 6 − 3 = 3

 **Solution to Question 74**

Maria needs 19 pineapples for making pineapples juice.
She already has 17 pineapples.
Number of pineapples she needs to buy more
= 19 − 17 = 2

 **Solution to Question 75**

Alvin has 20 balloons.
He has 9 balloons more than Ross.
= 20 − 9 = 11
Number of balloons Ross has = 11

224

Solution to Question 76

Karen planted 13 carrot seeds.
Only 5 of them sprouted.
Number of carrot seeds that did not
sprout = 13 − 5 = 8

Solution to Question 77

Fifteen children are attending a birthday party.
Nine of them are girls.
Number of boys attending the party = 15 − 9 = 6

Solution to Question 78

Farmer Jim had 18 chicken and ducks.
He had 10 chicken.
Number of ducks Jim had = 18 − 10 = 8

Solution to Question 79

The school is 13 kilometers away from Kim's home.
She has already walked 4 kilometers out of that.
Number of kilometers she still have to go to reach the school = 13
− 4 = 9 kilometers

Solution to Question 80

Mary has 45 metres of string.
She gave 10 meters from it to her daughter and 10 meters
to Grace. Total = 10 + 10 = 20 meters
Length of the remaining string = 45 − 20 = 25 meters

 **Solution to Question 81**

There are 30 roses in a vase.
7 of them are red, 4 of them are yellow and the rest are pink.
Total = 7 + 4 = 11
Number of pink roses in the vase = 30 − 11 = 19

 Solution to Question 82

Lucy baked a total of 35 cupcakes.
She baked 9 coffee cupcakes, some strawberry cupcakes and
18 banana cupcakes.
Total number of cupcakes = 9 + 18 = 27
Number of strawberry cupcakes she baked
= 35 − 27 = 8

 Solution to Question 83

Vanessa baked 19 lemon cakes and 26 chocolate cakes on
Monday.
On Tuesday, she baked 15 chocolate cakes.
Total number of chocolate cakes = 26 + 15 = 41
Number of chocolate cakes Vanessa baked more than
lemon cakes = 41 − 19 = 22

 Solution to Question 84

There are 21 boys and 12 girls in the table-tennis
club. There are 16 boys and 22 girls in the chess
club. Total number of boys = 21 + 16 = 37
Total number of girls = 12 + 22 = 34
Number of girls fewer than boys in both the clubs
= 37 − 34 = 3

 **Solution to Question 85**

There are 35 green candies and 12 red candies in a jar.
Total number of candies = 35 + 12 = 47
After Thomas ate some of the candies, there were 16 candies left.
Number of candies Thomas ate = 47 − 16 = 31

 Solution to Question 86

Jason has a collection of some stamps.
He received 16 stamps from his father and 12 stamps from his brother.
Total number stamps received = 16 + 12 = 28
He had 50 stamps in the end.
Number of stamps Jason had before receiving the stamps from his father and brother
= 50 − 28 = 22

 Solution to Question 87

There were 35 adults and 28 children at a party
Total members in the party = 35 + 28 = 63
4 men and 18 girls left the party.
Total number of people who left = 4 + 18 = 22
Number of people at the party in the end
= 63 − 22 = 41

 Solution to Question 88

Mrs. Watson had 32 eggs.
She threw away 10 rotten eggs and sold away 6 eggs.
= 10 + 6 = 16
She kept the rest = 32 − 16 = 16

Number of eggs Mrs Watson kept = 16

 Solution to Question **89**

There are 52 pages in a magazine.
Janice read 12 pages on Tuesday and 6 pages on Wednesday.
Total pages read = 12 + 6 = 18
Number of pages not read yet = 52 − 18 = 34

 Solution to Question **90**

There are 16 boys and 24 girls in a field.
Total number of students = 16 + 24 = 40
The P.E teacher has 40 balls and she wants to give two balls to each student .
Number of balls needed for 40 students = 40 + 40 = 80
Number of more balls needed = 80 − 40 = 40

 Solution to Question **91**

8 students went to the library.
2 of them did not borrow any books.
Children who borrowed books = 8 − 2 = 6
These 6 children borrowed 3 book each = 3 + 3 + 3 + 3 + 3 + 3 = 18
Number of books they borrowed altogether = 18

 Solution to Question **92**

James scored 40 points in a game.
Jason scored 8 points less than him = 40 − 8 = 32 Number of
points they scored altogether = 40 + 32 = 72

 ## Solution to Question 93

There are 28 paper clips in a box.
Kalvin has 2 such boxes
Total number of paper clips = 28 + 28 = 56
He used 8 paper clips and gave away 3 paper clips to his friend.
Number of used clips = 8 + 3 = 11 Number of paper clips left in
the box = 56 − 11 = 45

 ## Solution to Question 94

Jessica bought 5 packets of balloons and used 2 out of them.
Remaining number of balloon packets = 5 − 2 = 3
There were 3 balloons in each packet .
Number of balloons she had left = 3 + 3 + 3 = 9

 ## Solution to Question 95

Jason, Anna and Tom had 80 marbles. Jason had 25 marbles.
Anna had 12 marbles more than him = 25 + 12 = 37
Total number of marbles = 25 + 37 = 62 Number of marbles
Tom had = 80 − 62 = 18

 ## Solution to Question 96

There are 38 roses in a vase.
12 of them are red, 10 are pink and the rest are white.
Total number of roses = 12 + 10 = 22 Number of white roses
= 38 − 22 = 16

 **Solution to Question 97**

Mr. Smith wants to buy 60 fruits.
He bought 18 apples and 19 bananas from the super-market.
Total Number of fruits bought = 18 + 19 = 37
Number of fruits he still needs to buy = 60 − 37 = 23

 **Solution to Question 98**

Box A contains 52 cards. Box B contains twice as many cards as
Box A = 52 + 52 = 104
Number of cards in the two boxes = 52 + 104 = 156

 **Solution to Question 99**

There are 26 boys and 14 girls in a class.
Total number of children = 26 + 14 = 40
17 children wear spectacles.
Number of children who do not wear spectacles = 40 − 17 = 23

 **Solution to Question 100**

Christina had 80 cookies
She gave 7 cookies each to her 3 friends = 7 + 7 + 7 = 21

Number of cookies she has left = 80 − 21 = 59

 **Solution to Question 101**

Jane had 35 stickers.

She gave 12 stickers to Samuel.
Remaining stickers = 35 − 12 = 23
William gave 3 more stickers to Jane.
Number of stickers Jane had left = 23 + 3 = 26

 ### Solution to Question **102**

There are 45 books on a shelf.
12 of them are English books, 7 are science books and the rest are mathematics books.
Total number of books = 12 + 7 = 19
Number of mathematics books = 45 − 19 = 26

 ### Solution to Question **103**

There were 54 apples in a basket and 12 fewer oranges than apples in the basket.
Number of oranges = 54 − 12 = 42
Number of fruits altogether = 54 + 42 = 96

 ### Solution to Question **104**

There were 40 fish in a tank.
16 of them were orange, 11 were red and the rest were white.
Total = 16 + 11 = 27
Number of white fish in the tank = 40 − 27 = 13

 ### Solution to Question **105**

A zoo-keeper has to feed 2 bananas to every monkey . There were 12 monkeys.
Bananas needed for 12 monkeys = 2 + 2 + 2 + 2 + 2 + 2 + 2 + 2 + 2 + 2 +2 + 2 = 24

He had 20 bananas in his basket.
Number of more bananas needed to feed all the monkeys = 24 − 20
= 4

Solution to Question 106

There are 12 children altogether.

Solution to Question 107

Joel has a toy car collection.
His mother gave him 18 toy cars and his father gave him another 8 toy cars.
Total cars received = 18 + 8 = 26
He now has a total of 40 toy cars altogether.
Number of toy cars Joel had at first = 40 − 26 = 14

Solution to Question 108

Thomas had 28 toy soldiers.
12 of them were red, 5 were green and the rest were blue.
Total soldiers = 12 + 5 = 17
Blue toy soldiers = 28 − 17 = 11
Difference between the number of blue toy soldiers and green toy soldiers
 = 11 − 5 = 6
Thomas had 6 more blue toy soldiers than green toy soldiers.

Solution to Question 109

80 students participated in a chess competition.
15 of them were in Grade 2, 13 were in Grade 3, and the rest were in Grade 4.
Total = 15 + 13 = 28
Number of Grade 4 students = 80 − 28 = 52
Total number of Grade 4 and Grade 3 students who participated in the competition = 52 + 13 = 65

Solution to Question 110

There are 46 tulips in a vase.
There are 28 more tulips than roses in the vase Roses = 46 − 28 = 18
Total number of flowers in the vase = 18 + 46 = 64

Solution to Question 111

A photo album contained 25 photographs.
Jason took out 6 photographs = 25 − 6 = 19
He put in 12 new ones.
Number of photographs in the photo album now = 19 + 12 = 31

Solution to Question 112

Susan baked 7 strawberry cupcakes.
Claire baked 3 fewer cupcakes than Susan
= 7 − 3 = 4
Number of strawberry cupcakes they made altogether = 4 + 7 = 11

Solution to Question 113

Jason and Kelvin have 25 books.

Jason has 5 more books than Kelvin.

If you take away the difference of 5 books between Jason and Kelvin, it would leave them both with an equal number of books, i.e 10 books each.

But Jason has five more books, this means he has
= 10 + 5 = 15 books

Solution to Question 114

Larry has 16 oranges.

He gave 7 of them to his friend and gave 2 to his mother. Total number of oranges given away = 7 + 2 = 9

Number of oranges finally left with Larry = 16 − 9 = 7

Solution to Question 115

Samuel and Jason have a total of 40 fish. Samuel has 10 fish more than Jason.

If you take away the difference of 10 fish between Samuel and Jason, it would leave them both with an equal number of fish, i.e 15 fish each. But Samuel has 10 fish more than Jason, this means Samuel has = 15 + 10 = 25 fish

Solution to Question 116

Liz and Kevin have a total of 50 eggs.

Kevin has 10 eggs more than Liz.

Difference = 50 − 10 = 40

No. of eggs with Kevin = 20 +10 = 30

No. of eggs with Liz = 20
If they were to have the same number of eggs Kevin should give
5 eggs to Liz.

Solution to Question 117

Tom has 10 marbles.
James has 7 marbles lesser than Tom = 10 − 7 = 3
No. of marbles they have in all = 10 + 3 = 13

Solution to Question 118

Jake has 14 marbles.
He lost 4 marbles while playing = 14 − 4 = 10
He bought 6 marbles = 10 + 6 = 16
Number of marbles Jake has now = 16

Solution to Question 119

Alex has 200 balloons.3 of them are red, 45 of them are blue.
Number of yellow balloons = 200 - 48 = 152

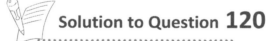

Solution to Question 120

There are a total of 100 students and teachers in a school.
30 of them are boys, 25 are girls and the rest are teachers.
Total number of students = 30 + 25 = 55
Number of teachers = 100 − 55 = 45
Number of teachers more than the number of girls = 45 − 25 = 20

Solution to Question **121**

Maya and her family went on a vacation for 14 days. They spent 5 days in Germany and 4 days in Italy. Total = 5 + 4 = 9 days
They spent rest of the time in France.
Number of days they spent in France = 14 − 9 = 5 days

 ### Solution to Question **122**

There are 30 books on a shelf.
10 of them are English books.
15 are science books.
Total = 10 + 15 = 25
The rest are mathematics books.
Number of mathematics books = 30 − 25 = 5

 ### Solution to Question **123**

Lucy baked 30 cupcakes.
She baked 10 coffee cupcakes and 10 banana cupcakes.
Total = 10 + 10 = 20
The rest are strawberry cupcakes.
Number of strawberry cupcakes Lucy baked
= 30 − 20 = 10

 ### Solution to Question **124**

Mum wants to give 20 oranges to Amie, Nathan, Mike and Daniel.
= 5 + 5 + 5 + 5 = 20
Number of oranges each child receives = 5

236

Solution to Question 125

On a farm, there are 5 sheep, 5 ducks, 2 cows, 2 cats and the farmer.
Sheep = 4 legs = 4 + 4 + 4 + 4 + 4 = 20
Ducks = 2 legs = 2 + 2 + 2 + 2 + 2 = 10
Cows = 4 legs = 4 + 4 = 8
Cats = 4 legs = 4 + 4 = 8
Farmer = 2 legs
Total number of legs = 20 + 10 + 8 + 8 + 2 = 48

Solution to Question 126

Alice put 8 stamps on every page of her stamp album.
There were 5 pages in her album.
Number of stamps she had altogether
= 8 + 8 + 8 + 8 + 8 = 40

Solution to Question 127

There were 20 marbles in a bag.
Sam divided them equally into 4 groups
= 5 + 5 + 5 + 5 = 20
Number of marbles in each group = 5
Number of marbles in 2 such groups = 5 + 5 = 10

Solution to Question 128

There were 12 chicken eggs and 6 duck eggs in a basket
Total eggs = 12 + 6 = 18
All the eggs were packed equally into 3 boxes = 6 + 6 + 6.
Number of eggs in each box = 6

Solution to Question 129

Bryan had 4 bags.
There were 9 oranges in each bag.
Total number of oranges = 9 + 9 + 9 + 9 = 36
He gave 12 oranges to his friend.
Number of oranges he had left
= 36 − 12 = 24

 Solution to Question 130

Richard had 3 boxes of cookies.
There were 5 cookies in each box.
Total number of cookies = 5 + 5 + 5 = 15
His friend gave him another 12 cookies.
Number of cookies Richard has altogether = 15 + 12
= 27

 Solution to Question 131

Karen packed 12 buns into 3 packets
= 4 + 4 + 4 = 12
Number of buns in each packet = 4
In each packet, there was 1 raisin bun and some butter buns.
Number of butter buns in each packet = 4 − 1 = 3

 Solution to Question 132

There were 6 flowers in a bouquet. Dorothy bought 3
such bouquets.
Total number of flowers = 6 + 6 + 6 = 18
Number of flowers she bought altogether = 18

238

Solution to Question **133**

There are 4 bags of oranges.
Each bag has 9 oranges.
Total number of oranges = 9 + 9 + 9 + 9 = 36
There are 3 bags of mangoes.
Each bag has 6 mangoes.
Total number of mangoes = 6 + 6 + 6 = 18
Number of fruits altogether = 36 + 18 = 54

 ## Solution to Question **134**

16 buns are shared equally among 4 children
= 4 + 4 + 4 + 4 = 16
Number of buns each child gets = 4
Each child eats 2 buns.
Number of buns left per child = 4 − 2 = 2
Number of buns left with 4 children in the end = 2 + 2 + 2 + 2 = 8

 ## Solution to Question **135**

2 books cost $10 = 5 + 5
Cost of 1 book=5
 4 pens cost $24 = 6 + 6 + 6 + 6
Cost of 1 pen = $6
Amount of money Henry paid when he bought one book and one pen = $5 + $6 = $11

 Solution to Question **136**

A fruit seller has 8 bunches of bananas.
Each bunch has 4 bananas.
Total number of bananas = 4 + 4 + 4 + 4 + 4 + 4 + 4 + 4 = 32
23 of the bananas are rotten, and the rest of the bananas get sold.
Number of bananas sold by the fruit seller = 32 − 23 = 9

 Solution to Question **137**

Kelvin has a total of 16 potatoes.
He wants to put 4 potatoes in one bag.
= 4 + 4 + 4 + 4 = 16
Number of bags he needs to pack all the potatoes = 4

 Solution to Question **138**

A teacher asker her students to stand in 3 rows that had 7 students each.
Total number of children in 3 rows = 7 + 7 + 7 = 21
After some time 2 new students were added in each row.
Total number of children altogether now = 21 + 6 = 27

 Solution to Question **139**

Thomas had 10 boxes of egg tarts.
He gave 6 boxes of egg tarts to his friends. Remaining number of boxes = 10 − 6 = 4
There were 4 egg tarts in each box.
Number of egg tarts left with Thomas = 4 + 4 + 4 + 4 = 16

Solution to Question 140

Larry bought 30 lemons.
He wants to pack all the lemons into packets of 5
= 5 + 5 + 5 + 5 + 5 + 5 = 30
Number of packets he would need for packing all the lemons
= 6

Solution to Question 141

Richard has 35 bananas.
He wants to pack 5 bananas into one packet. = 5 + 5 + 5 + 5 + 5
+ 5 + 5 = 35
a) Number of packets Richard packed = 7
b) Each packet of banana was sold at $2.
 Money he earned after selling all the bananas = 2 + 2 + 2 + 2 +
2 + 2 + 2 = $14

Solution to Question 142

Jeffrey has 2 mangoes.
Harry has twice the amount of mangoes than Jeffrey = 2 + 2 = 4
Victor has thrice the amount of mangoes than Jeffrey = 2 + 2 + 2 = 6
Number of mangoes that the 3 children have altogether = 2 + 4 + 6 = 12

Solution to Question 143

A chicken has 2 legs, and a dog has 4 legs. There were
4 chicken and 2 dogs on the farm.
Number of legs for 4 chicken = 2 + 2 + 2 +2 = 8
Number of legs for 2 dogs = 4 + 4 = 8
Total number of legs = 8 + 8 = 16

 ## Solution to Question **144**

There were 4 pigs and some birds on the farm.
There were a total of 20 legs.
Pigs have 4 legs
4 pigs = 4 + 4 + 4+ 4 = 16
Total number of bird legs = 20 − 16 = 4
A Bird has 2 legs.
Number of birds on the farm = 2

 ## Solution to Question **145**

Ella has 15 stickers.
She gives 3 stickers to each of her 3 sisters = 3 + 3 + 3 = 9
Number of stickers she had left = 15 − 9 = 6

 ## Solution to Question **146**

Lena has 15 cookies.
She packs 5 cookies in each box.
= 5 + 5 + 5 = 15
Number of boxes she needs to pack 15 cookies = 3

 ## Solution to Question **147**

Tom has 3 cards now.
He buys 3 new cards every day.
3 + 3 + 3 + 3 + 3 + 3 = 18
Number of days he takes to collect 18 cards altogether = 5
(since he already had 3 cards in the beginning)

Solution to Question 148

One sparrow has 2 legs
Number of feet on the branch = 2 + 2 + 2 + 2 = 8

Solution to Question 149

Mum baked some cupcakes for Tim, Liz and Jill.
Each of them ate 2 and none were left.
Mum baked = 2 + 2 + 2 = 6 cupcakes

Solution to Question 150

Alice drew 2 flowers on every page of her drawing book.
There were 10 pages in her drawing book.
= 2 + 2 + 2+ 2+ 2 + 2 + 2+ 2+ 2 = 20
Number of flowers she drew altogether = 20

Solution to Question 151

Tina's birthday is on November 21st. Lucy's
birthday is a week before that.
Lucy's birthday = 21 − 7 = 14th November

Solution to Question 152

Number of days left for your birthday to come = 21 − 4
= 17 days

Solution to Question 153

1. Eat snack. 2. Do homework.

3. Ben plays outside at 3:00.

Solution to Question 154

Eva is 4 years old and Jane is 8.
Number of years in which Jane will be 11 = 11 − 8 = 3 years
After 3 years Eva will be = 4 + 3 = 7 years.

Solution to Question 155

The sum of Kelvin and his father's age is 50 years.
His father is 40 years older than him.
Difference between their ages = 50 − 40 = 10 years
Kelvin = 5 years old
Father = 5 + 40 = 45 years old

Solution to Question 156

Bill is 3 years old.
Ann is twice as old as Bill = 3 + 3 = 6 years
Bill is three times as old as Carol.
= 1 + 1 + 1 = 3

244

Therefore Carol = 1 year
Ages of 3 children:
Bill = 3 years
Ann = 6 years
Carol = 1 year

Solution to Question 157

Andy reads two pages every five minutes.
20 min = 5 min + 5 min + 5 min+ 5 min
Number of pages Andy will read after twenty minutes =
2 + 2 + 2+ 2 = 8

Solution to Question 158

Kevin wakes up at 6 o'clock every day.
He slept at 10 PM in the night.
Number of hours he slept = 8 hours

Solution to Question 159

A movie started at 12'o clock and ended after 3 hours.
The time at which the movie ended
= 12 + 3 = 3 o' clock

Solution to Question 160

Carl attended a birthday party.
The party started at 2.00 P.M. and ended at 4.00 P.M.
Number of hours the birthday party lasted
= 4 − 2 = 2 hours

Solution to Question **161**

Tom left for work at 8:00 A.M..
He got stuck in traffic for 2 hours later. Tom reached at work at
= 8 + 2 = 10 A.M..

 ## Solution to Question **162**

Mike does his homework one hour after dinner.
He eats dinner at 6:30 P.M.
He starts doing his homework at = 6:30 + 1 hour = 7:30 P.M.

 ## Solution to Question **163**

Lucy started baking a cake at 1:00 P.M.
It should bake for 3 hours.
The cake be would be ready at 1:00 P.M. + 3 hours = 4 P.M.

 ## Solution to Question **164**

I went cycling to the park yesterday.
It takes me 3 hours to get there.
I start at 9:00 A.M.
Time at which I will reach the park = 9:00 A.M. + 3 hours = 12
P.M.

 ## Solution to Question **165**

Danny has a total of 6 apples and oranges.
There were 2 more oranges than apples. 2
apples + 4 oranges = 6
He sold oranges at a price of $1.
a) Oranges he had = 4

246

b) Money earned by selling oranges = $1 + $1 + $1 + $1 = $ 4

 ## Solution to Question 166

Amelia and Raymond together have $10.
Raymond gave $2 to Amelia and they now have same amount of money.
If both of them have the same amount of money, they have $5 each.
But Amelia has $5 after Raymond gave her $2. This means in the beginning Amelia had= $5 − $2 = $3
Amount of money Raymond had in the beginning = $10 − $3 = 7$

 ## Solution to Question 167

Peter had $20.
He spent $4 on Monday and $8 on Tuesday.
Total money spent = $4 + $8 = 12
Money left = $20 − $12 = $8

 ## Solution to Question 168

A pencil cost 40 cents and an eraser cost 20 cents.
Jim has 80 cents to spend.
He bought a pencil. Therefore remaining amount = 80 − 40 = 40 cents
Each eraser costs 20 cents = 20 cent + 20 cent = 40 cent
Number of erasers Jim could buy = 2

 ## Solution to Question 169

A watch and a clock cost $59 in a sale.
The watch cost $39.

Cost of the clock = $59 − $39 = $20
Difference between the cost of watch and the clock = $39 − $20 = $19
The watch cost $19 more than the clock.

 ## Solution to Question **170**

Peter had 90 cents in his wallet.
He had 5 ten-cent coins and some twenty-cent coins.
Amount of 10 cents coins = 10 + 10 + 10 + 10 + 10 = 50 cents
Remaining amount = 90 − 50 = 40 cents
Twenty coins = 20 + 20 = 40
Number of twenty-cent coins he had = 2

 ## Solution to Question **171**

A shirt cost $41.
A pair of pants cost $12 more than the shirt = $41 + $12 = $53
Cost of shirt and the pair of pants together = $41 + $53 = $94

 ## Solution to Question **172**

Andy spent $72 on a belt and saved the rest of his money. He saved $12,
Amount of money he had at first = $72 + $12 = $84
Amount of money he spent more than he saved = $72 − $12 = $60

 ## Solution to Quest on **173**

Maggie had $15.
She bought a chocolate cake and had $7 left.
Cost of chocolate cake = $15 − $7 = $8
2 such chocolate cakes cost = $8 + $8 = $16

248

Solution to Question 174

Jason had $80.
Linda had $36.
Difference between amount of money Jason has than Linda
= $80 − $36 = $44

 Solution to Quest on 175

A magazine costs $9.
Abook cost $5 more than the magazine
= $9 + $5 = $14
Cost of 2 such books = $14 + $14= $28

 Solution to Question 176

Lawrence had $20.
He bought a vase for $12.
Money left with Lawrence = $20 − $12 = $8
The cashier gave him a five dollar note and some two-dollar notes.
Amount of $2 notes = $8 − $5 = 3$
Number of $2 notes = 1

 Solution to Question 177

A stalk of rose cost $5.
A florist sold 3 stalks of roses and some stalks of orchids.
Cost of 3 stalks of rose = 5 + 5 + 5 = $15
She earned $27.
Money she earned from selling all the orchids = $27 − $15 = $12

Solution to Question 178

A dinner cost $100.
Ethan paid $35, Marcus paid $40 and Leonard paid the rest.
Total = $35 + $40 = $75
Amount of money Leonard paid
= $100 − $75 = $25

 Solution to Question 179

A plate cost $23.
The plate costs $8 more than a bowl.
Cost of Bowl = $23 − $8 = $15
Total cost of the plate and the bowl = $23 + $15 = $38

 Solution to Question 180

A chair costs $8.
Cost of 2 chairs = 8 + 8 = $16
2 chairs and a table cost $50.
Cost of the table = $50 − $16 = $34

 Solution to Question 181

Maggie bought a electric oven for $59.
She gave the cashier $100.
Change she got back = $100 − $59 = $41

 Solution to Question 182

Jason had $40.

250

He spent $6 on food, $9 on books and some on stationery
Total money spent = $6 + $9 = $15
He had $8 left
Amount he spent on stationery
= $40 − $15 − $8 = $17

 ### Solution to Question 183

John had $8.
Doris had $3 more than John.
= $8 + $3 = $11
Paul had $5 less than Doris = $11 − $5 = $6
Amount of money they have altogether = $8 + $11 + $6 = $25

 ### Solution to Question 184

An orange costs 50 cents.
A pears cost 20 cents less than an orange.
= 50 − 20 = 30 cents
A mango cost 60 cents more than the pear = 30 + 60 = 90 cents
Cost of the pear = 30 cents

 ### Solution to Question 185

A dress costs $28.
It costs $18 more than a blouse.
Blouse = $28 − $18 = $10
The total cost of the dress and blouse
= $10 + $28 = $38

Solution to Question 186

Richard had $100.
He spent $26 and saved the rest of his money.
Amount of money he saved
= $100 − $26 = $74

 Solution to Question 187

There are 3 roses in a bunch.
Each bunch cost $4. Caroline had $20.
= 4 + 4 + 4 + 4 + 4 = 20
Number of bunches she can buy = 5
Number of roses she bought altogether = 3 + 3 + 3 + 3 + 3 =
15

 Solution to Question 188

Henry bought some biscuits that cost $6.
He bought a drink at $8.
Total money spent by Henry = $ 8 + $6 = $14

 Solution to Question 189

Larry has 3 ten-dollar notes
= $10 + $10 + $10 = $30
He bought 4 cupcakes at a coffee shop.
Each cupcake cost $5.
Money Spent = $5 + $5 + $5 + $5 = $20
Change he received from the cashier
= $30 − $20 = $10

 **Solution to Question 190**

John saves $2 every day.
Kelvin saves $1 less than John every day
= 2 − 1 = $1
John wants to save a total of
$14 = 2 + 2 + 2 + 2 + 2 + 2 + 2
Number of days in which John would save $14 = 7
Number of days in which John would save $16 = 8
Amount of money Kelvin will save by the time John saves $16 =
1 + 1 + 1 + 1 + 1 + 1 + 1 + 1 = $8

 Solution to Question 191

Gavin went to a shop with a $10 note.
He bought 2 packets of potato chips and 2 bottles of Coca-Cola.
Each packet of potato chip cost $2.
Each bottle of Coca-Cola cost $1.
2 packets cost = 2 + 2 = $4
2 bottles cost = 1 + 1 = $2
Total money spent = $4 + $2 = $6
Change he received from the cashier
= $10 − $6 = $4

 Solution to Question 192

Jenny went to the market with 1 fifty dollar note = $50
She bought 2 fish at $5 each = $5 + $5 = $10
She bought 3 watermelons at $10 each = $10 + $10 + $10 = $30 Total
money spent = $30 + $10 = $40
Change she received from the cashier
= $50 − $40 = $10

Solution to Question **193**

Kevin bought 6 pairs of socks.
Each pair of socks cost $2.
Cost = 2 + 2 + 2 + 2+ 2 +2 = $12
He gave $20 to the cashier.
Amount of money he spent in all = $12
Change he received from the cashier
= $20 − $12 = $8

Solution to Question **194**

Ben had three $10 notes
= $10 + $10 + $10 = $30
and two $ 5 notes with him = $5+ $5 = $10 His father
gave him one $2 note.
Total amount of money Ben had
= $30 + $10 + $2 = $42
Using all his money Ben bought a game for $ 40.
Amount of money left with him
= $42 − $40 = $2

Solution to Question **195**

Tim paid $5 for a chocolate.
He received two 50 cent coins in change
= 50 cents + 50 cents = $1
Cost of the chocolate = $5 − $1 = $4

Solution to Question **196**

I want to buy a pencil that costs 50 cents.
I pay a 2 dollar note to the shopkeeper= 50 cents + 50 cents + 50

cents + 50 cents = $2
Change that I will get back = 50 + 50 +50 = 150 cents = $1 and 50 cents

 ## Solution to Question **197**

Weight of fruits + basket = 1 + 4 + 2 = 7 kg
Weight of basket = 1 kg
Therefore weight of the fruits
= 7 − 1 = 6 kg

 ## Solution to Question **198**

The temperature at noon today is 40 degree.
The temperature dropped by 10 degrees in the evening.
Temperature now = 40 − 10 = 30 degrees

 ## Solution to Question **199**

I measured my pencil at 9 inches.
My friend's pencil is 16 inches.
Number of inches my friend's pencil is longer than mine = 16 − 9
= 7 inches

 ## Solution to Question **200**

My study table is 5 feet wide.
The room's door is 3 feet wide.
Number of feet by which my room's door needs to be wider to fit my desk = 5 − 3 = 2 feet.